In Case Of
Nuclear Attack

EMERGENCY HANDBOOK
for the Unprepared with
Advice on being Best Prepared

In Case of Nuclear Attack Emergency Handbook
Copyright 2023 George Gibson
Hunker Down Publishers
ISBN: 9798372745315
Midtown New York, NY
Cover graphic by Catalania at Pixabay

Name	
Address	
Contact	
Emergency contacts	
Family and close contacts	
2-way radio frequency / channel*	_____ AM ☐ FM ☐ LSB ☐ USB ☐

*Nominate frequencies that you and loved ones will use on two-way radio transceivers such as CB radio.

Community shelter	
Shelter Address	
Community shelter	
Shelter Address	
Community shelter	
Shelter Address	

Emergency Checklist

*Refer to preparation tables at the end of this book.
Check them off here and add any other items.*

☐ Food	☐ Whistle(s)	☐
☐ Water	☐ Radios	☐
☐ Flashlights	☐ 2-way radios	☐
☐ Batteries	☐ Plastic bags	☐
☐ Blankets	☐ Toolbox	☐
☐ First Aid	☐ Rain shield	☐
☐ Phones	☐ Jerry cans	☐
☐ Chargers	☐ Gasoline	☐
☐ Tools	☐ Firearms	☐
☐ Duct tape	☐ Geiger counter	☐
☐ Masks	☐ Dosimeter	☐
☐ Goggles	☐ Packed soil	☐
☐ Medications	☐ Earth filters	☐
☐ Generator	☐ Air filters	☐

No matter how well you prepare for a nuclear cataclysm, you could be caught outside away from your shelter. You could be out driving, or shopping, or on the street.

This quick guide will help you learn practical steps in best protecting yourself against an impending nuclear strike no matter where you are.

- Read this book carefully and commit to memory.
- Keep this book in a safe, handy place such as a first aid kit.
- Keep a spare in your vehicle's glove compartment.

This book has three sections for the unprepared outlining procedures *before the blast*, *during the blast*, and *after the blast*. A fourth section provides the preparation guide, **Best to Prepare**, should you be lucky to have time to prepare for a possible future calamity.

SECTION 1

Before The Blast

You have just heard news of an impending nuclear strike...

You have less than 30 minutes

...to minimize the effects of a blast that includes:

- EMP (Electro-Magnetic Pulse) that can knock out electronics, including electronics in your vehicle

- Impact flash that can cause blindness

- Radiation that can destroy cells

- Fire and heat that can kill and cause injury

- Radioactive fallout from dirt, debris and tainted rainwater

What you do in the next 30 minutes is crucial to staying alive...

Keep listening for news

Listen for emergency broadcasts. They typically override regular television and radio programs. Listen carefully for instructions. You may also receive emergency alert text messages on your phone and hear an air raid siren.

- Find a portable radio to keep with you; a battery-powered or hand cranked type is recommended in case of a power outage.

- Keep your phone with you at all times for emergency text message updates. Keep your charger with you.

- Keep bottled water and snack bars with you and wear sturdy footwear.

- Go straight to your basement or find a shelter elsewhere as soon as possible. If there is no shelter nearby or you are near the city, drive as far away as possible and seek cover in the basement and/or center of a solid building having few windows.

- If you live in a town that is known to have silos or military bases, leave as soon as you can. Missiles that are aimed at military sites or ports and factories would have maximum yield, so make sure you are not outside.

- If you have a handy talkie, take that with you wherever you go. CB radios are useful in emergencies along with 24 hour emergency weather radios.

The following pages show you what to do if you are:

- ▶ *in your vehicle*
- ▶ *stuck in traffic or public transit*
- ▶ *at school or college*
- ▶ *at the office*
- ▶ *at home*

If you are in your vehicle

EMP from an initial blast may damage the electronics in your car making it unusable. That's why you must drive as far away from a large populated area and to a shelter as soon as you hear of an impending strike.

- Drive home, as your home will offer some safety over being outside, especially if you have a shelter prepared.

  This will depend on your proximity to the city and whether your home can offer enough protection such as a basement. If you are closer than 25 miles to a populated area, consider driving further away from the city and finding a better shelter.

- If you cannot find a shelter, find a basement parking lot in a building away from the city, drive to the center of the building and stay in your car. Only do this if you are confident the building is far enough away from the blast zone, beyond 25 miles – preferably further away, and will withstand a shock wave.

- If you are outside with nowhere safer and only minutes to spare, pull over, place a sun shade across your windscreen (if you have one) and try to block the other windows with anything that you may have in your car such as a coat or jacket. Adjust your seat back, face downwards, cover your eyes with the fold of your arm and wait for the flash and shockwave to pass.

- Prior to a blast, turn *off* your two-way and broadcast radios as this may help protect against EMP.

- DO NOT OPEN YOUR EYES DURING THE FLASH. You could be blinded.

- After the blast find a better place to shelter as *you only have 15 minutes before radioactive fallout* reaches the ground.

- Do not stay outside. Cars, trucks and trailers are not suitable fallout shelters.

- Find shelter in the opposite direction of any fallen buildings, away from the wind and as far away from the blast.

- Look for 'Fallout Shelter' signs with radiation symbols on any buildings nearby. If there are none, find a solid brick or concrete building with few windows where you can head for the basement to avoid the descending cloud of radiation.

    If there is no basement choose an area in the middle of the building away from windows, to avoid shattering glass.

- Always look for concrete or brick buildings. If there are no strong buildings nearby, seek indoors anywhere you can find and avoid being near windows.

- The John Hopkins Center for Health Security (CHS) in Baltimore states that radiation exposure decreases by 55% an hour after an explosion and 80% after 24 hours.

Therefore you must stay in the shelter for more than 24 hours.

48 hours is recommended.

If the shelter you are in has ample supplies of food, water, and clothing, plan to stay for a minimum of two weeks and do not venture outside.

If you are stuck in traffic, bus, or train

Trying to leave the city will be on the minds of most people. If you happen to be caught in a transport jam you will likely be exposed outside:

- Stop your engine and apply the handbrake.

- Prior to a blast, turn *off* your two-way and broadcast radios as this may help protect against EMP.

- Find the nearest building, bridge or overpass in which to shelter.

- If you cannot leave your car, place a sun shade across your windscreen if you have one and try to block the other windows with anything you may have in your car such as a coat or jacket. Place your seat back, face downwards, cover your eyes with the fold of your arm and wait for the flash and shockwave to pass.

- DO NOT OPEN YOUR EYES DURING THE FLASH. You could be blinded.

- After the blast find a better place to shelter as *you only have 15 minutes before radioactive fallout* reaches the ground.

Since you are in traffic there will be many people who are panicking.

- If you can avoid others do so. However it is prudent to ask others if they know of a shelter nearby.

- If you have your handy talkie transceivers, take them with you (even if you hear nothing on them after an EMP blast as the metal body of your car may have provided some protection).

- If you are on a bus, the driver may know of a safe shelter nearby. Do not panic. Leave the bus in an orderly fashion.

- If you happen upon a shopping center where looting is occurring, stay well away.

- If you are on a train get off at the next stop and STAY WHERE YOU ARE. You probably have little time to find a shelter elsewhere. Most city subway stations offer reasonable shelter.

    Go to the lower subway levels and stay for a minimum of 48 hours.

- If your train's power is cut and you are in a subway tunnel a long way from an exit, it's probably a better place to be after a blast. Stay for a minimum of 48 hours after the blast.

- If there is no power, try using your phone's flashlight, if it still works.

- When it is safe to move, head to the nearest underground platform as the cellphone network may still work in some areas.

    Note: There are purpose-built subways that are designed to withstand a nuclear blast (such as Moscow's subway system).

If you are a regular transit commuter, check with your state's subway system information and note below.

My local transit station shelter:

_____

_____

_____

_____

If you are at school

All schools, colleges and universities should have a procedure in place that includes the provision of water and food, along with two-way radios and broadcast radios.

There will probably be a Geiger counter at the science lab.

A public address broadcast will be made across campus. Follow instructions carefully.

- Go straight to the closest nominated fallout shelter and await further instructions.

- If there is no shelter, go to your nominated assembly area that is away from windows and preferably in a basement or sub-basement area.

- Fill water tubs and bottles.

- Shut off ventilation systems and air conditioning.

- Unplug all your wired devices such as computers and monitors to reduce risk of EMP damage. If you have metal desks, place them underneath, or metal lockers place them inside.

- Teachers should remove food from vending machines and take with them to the shelter to pass around the students when necessary.

    Remember, you must *remain in the shelter for a minimum of 48 hours*.

    If the shelter you are in has ample supplies of food, water, and clothing, plan to stay for a minimum of two weeks and do not venture outside.

- Practice social distancing; try not to huddle, especially after a blast has occurred.

If you are at work

Traffic chaos could jam the roads and you may not be able to leave for home in time.

Observe any company announcements on emergency procedures.

- Stay away from windows.

- Shut off ventilation systems and air conditioning.

- Unplug all your wired devices such as computers and monitors to reduce risk of EMP damage. Place under your desk or in your locker if time permits.

- Managers should remove food from vending machines and take with them to the shelter to pass around the staff when necessary.

- Use the fire stairs to lower levels.

- Do NOT use the elevators as EMP may cut electricity which could result in you being stuck in the elevator.

- Move to the center of the building on the lower floors, or basement, or parking lot.

- Remember, *stay in the shelter for more than 24 hours.*

    48 hours is recommended.

    If the shelter you are in has ample supplies of food, water, and clothing, plan to stay for a minimum of two weeks and do not venture outside.

If you are home

Although you may be prepared with a basement or dedicated shelter with water filters and/or bottled water supplies, food stock, air filtration, wind-up radio, flashlights, CB transceiver, and so on, there are still things you can do to your home that will reduce the impact of fallout.

For prepared or unprepared:

- Water supplies may be cut off or tainted after a blast. It is essential you have enough drinking water. Fill up all the buckets you have.

- Fill up your bathtub(s) with cold water.

- If you have time to clean your trash cans or bins, they would hold a lot of water.

- Shut off ventilation systems and air conditioning.

- Unplug all your wired home devices such as televisions, computers and network devices, to reduce risk of EMP damage.

- Use gaffer or duct tape or a similar strong thick tape and seal the gaps in your windows and doors. This will aid in the event of blast debris and dust.

- Run tape across your window glass;

    - [+] Make a cross; top edge middle to bottom edge middle, and left edge middle to right edge middle, and then;

    - [X] Make an 'X'; diagonally from each corner to help hold the glass in place.

- You may not have time to cover all windows so apply tape to the larger ones first.

- The blast can still shatter your windows depending on impact distance. Radioactive fallout can land on the roof and walls of your house so it is essential to stay in the middle of your house and preferably the basement.

- Close all blinds and turn off all air conditioners and fans.

- Block the chimney / fireplace using any sheet metal you can find and secure with duct or gaffer tape. Sheet metal will inhibit radiation penetration better than wood.

  Bags of packed earth a few feet thick will block much of the rays. You could place these in the fire place and block the chimney completely.

  Approximately every five inches of packed earth will halve the amount of gamma radiation reaching you.

- Make sure your pets are inside.

- You must consider the emotional wellbeing of yourself, your partner and children. Keep the exposure of bad media reports to a minimum when children are nearby because kids will see such things as very scary.

- Share food and water with others and keep to a minimum if there are more people seeking shelter at your place.

If you live in an apartment you should have a safety plan in place for the entire estate.

- Go to your designated assembly area. This would likely be the basement and/or parking lot area. Take food and water with you plus radios, cellphones, flashlights, warm clothes and blankets and good shoes.

SECTION 2

The blast has just occurred

Don't look up...

During the blast

Do not look in the blast direction as you will be blinded.

- You could be blinded up to 13 miles away (20.9 km) on a clear day and up to 53 miles (85.3 km) on a clear night.

- If you are outside you need to turn away from the blast, drop to the ground and cover your eyes from the flash with the fold of your arm. As the light dims, lay face down with your hands under your body so they do not get injured from the impending blast pressure and debris. You could also crouch under a desk or bed. The old 1950s 'Duck and Cover' slogan has some merit.

- Keep your mouth open as the initial pressure from the blast could burst your eardrums.

SECTION 3

After The Blast

You have 15 minutes...

The fallout will reach you within 15 minutes

After a nuclear explosion, the first few minutes are vital for survival, especially if you are outside.

- You have 15 minutes before the grains of fallout start to reach you.

- Exposure to these particles can result in radiation poisoning.

- Cover your mouth and nose if you are outside. Always have a mask available.

- Find shelter immediately.

In the home or shelter...

- Change your clothes if you have been exposed, and wash the exposed parts of your body (face, arms, hands, feet etc.).

- Carefully wash your hair and skin and use plenty of shampoo and soap. Try not to scratch yourself, try not to hurry. Don't use hair conditioner or skin moisturizer as this can bind with radioactive dust.

- Place removed clothes in a bag, seal it, and keep well away from others.

- If you are with others in a shelter, keep a social distance of six feet (around two meters) from others.

  Make sure those seeking shelter have removed the outer part of their clothes (jackets etc.) that may have been exposed.

- Ensure pets do not find a way outside. If they have been outside, brush their coats and then wash them carefully, and place towels in a separate bag, kept away from others.

- ***WAIT at least 48 hours*** before going outside. If you are at home with some provisions, plan to hunker down for a minimum of *two weeks*.

- When you go outside wear a mask and goggles if you have them, otherwise cover your face with cloth which should be discarded after use.

- The important thing in a nuclear aftermath is community. Banding together for common survival is the only way to move on.

  - Seek contact with those in authority.

  - Always check the identity of persons claiming to work in your best interest.

  - Identify sources of untainted food and water.

  - Avoid hospitals if you are not injured or affected by radiation.

SECTION 4

Best to Prepare

You don't know when the big one will drop...

Preparation Tables

In this world of increasing danger, it's wise to prepare as best you can within your budget. The following tables list recommended action items for:

- Home
- Your vehicle
- Workplace
- School

No doubt you'll think of more things you may need to round out the best possible survival result for you and your family.

At the front of this book is a checklist. Tick what you have already, and note the tables on the following pages to guide you to add any other items you think you will need.

| *Be discreet* | *Do not inform others of your preparations.* |

Preparation Tips for the Home

First aid	Have a substantial First Aid kit ready.
Food and water Don't forget toothpaste Don't forget Swiss Army knife; have more than two Consider building an under-house filtered rain-water tank	Have at least three months of non-perishable food and bottled water. Build up large water containers; one gallon per person per day. Bleach and potassium iodide will help to purify tainted water. There are many sites online explaining how to do this. Carbohydrates are recommended; stored in a cool, dry place such as a basement. Choose canned vegetables, packaged dried fruits, white rice, wheat, beans, sugar, honey, oats, pasta and long life or powdered milk. Make sure you have a spare can opener.

Two-way Radios CB mobile CB base	Have a pair of handy talkies. Also install a CB radio transceiver as a base radio. Use a 13.8 volt power supply with a 12 volt backup battery such as a car battery. Use a good quality indoor antenna to put up quickly in an emergency (so you don't have to go outside). When preparing, make sure you have at least 30 feet of RG58 coax cable (should you need to make a long run to the antenna, and an SWR (Standing Wave Ratio) meter to check the antenna match to radio (otherwise radio frequency power could damage your CB). Adjust your antenna's length to tune. If trapped inside, you could use an antenna matcher that will align your antenna to your radio if the antenna is compromised (off frequency).

Ham radio	Consider a ham radio technician certificate. You can talk longer distances.
NOAA weather radio	Use for 24 hour emergency broadcasts. Cranking type preferred.
Broadcast radio	Wind-up crank radios that don't rely on batteries are ideal in an emergency.
Spare lithium batteries	Include portable solar cells to charge them with. Larger current-rated batteries will last longer.
Metal box Faraday cage	Get a well-rated metal box to help protect your radios from EMP. Metallic space blankets may also reduce conductive voltages that may damage your radios. You can also wrap items in paper and then wrap that in aluminum foil.

Dust mask and goggles	In case you need to go outside, a mask and goggles will help protect you.
Toolbox containing wrench, pliers etc.	You may need tools to shut off water or gas mains and make any repairs around the house.
Whistles or panic alarms	Get one for each family member. Be careful when using near strangers or when you are in unfamiliar territory.
Survival kit	Make sure you have a survival kit! More info can be found here, https://www.redcross.org/get-help/how-to-prepare-for-emergencies/survival-kit-supplies.html
Block-out roller window shades	These are external metal weather-resistant window coverings. Make sure when installing you have a manual override control in case power is cut.

Generator	Have a generator and plenty of fuel to run it. Keep in a place where it is less likely to be stolen, such as your garage, and run fumes out the window via an exhaust conduit.
Flashlights	Also keep portable LED lamps at hand and solar chargers too.
Solar panels and batteries	Rooftop solar along with backup batteries. Foot generators may be of use during a nuclear winter.
Filtration units	Purify air coming into the house or shelter, and use filtered shower and tap heads. Consult with provider on specifics of filtration for the various types of radiation.
Firearms and knives	Provided you have a licensee to use weapons (check your state laws), keep on hand *but use as a last resort*.

Sealable Trash bags	These are used to place any exposed clothing or other radioactive material.
'Mud room' entry way	These are self-contained separate entry areas to the main building or shelter, similar to a landing or small foyer. The best 'mud room' is one having a shower with removable shower head and flexible hose, plus a clothes chute. Here you can also clean your pets. Ensure shower runoff leads directly to the outside and not across and under the house. Treat the room as you would an air lock.
Goggles and masks	Use eyewear that completely covers your face with no gaps, yet allows you to have the best view possible.
Duct tape	You can buy nuclear grade duct tape which can handle higher temperatures.

Cellphones	Some services may still operate, so keep your phones and chargers handy. Solar chargers are available for cellphones.
Medication	Keep your meds handy for at least a month's supply. Keep your medicine requirements history / info with you.
Gasoline	Have jerry cans filled and ready to go. Use and refill every month so the fuel is fresh.
Shelters	Consider building a sub-earth-packed shelter. Shelters built into and under packed soil are best at minimizing pressure waves from an explosion and they help to block gamma radiation.
Prepare a night kit	Have a bag full of essentials next to your bed as nukes could strike at night!

Dosimeters and Geiger counters, or a KFM meter	Dosimeters let you know how much radiation you have absorbed. Geiger counters are good but require batteries (usually D-cells). Kearny Fallout Meters do not need batteries to run. They don't need recalibration and are EMP-proof. They are easy to build and instructions can be found online.
Iodine tablets*	Discuss with your doctor if you need to have these as there are risks involved.

* It is not recommended to take iodine tablets unless authorized by a professional.

Follow the CDC guideline here,
https://www.cdc.gov/nceh/radiation/emergencies/ki.htm

Fact sheet on usage:
https://www.health.ny.gov/environmental/radiological/potassium_iodide/fact_sheet.htm

Preparation Tips for Your Vehicles

Water bottles	Always keep one or more in the vehicle.
Food	Chocolate bar or other non-perishable. Replenish when used.
CB radio	Keep a handheld radio in the glove compartment. Also use a mobile CB with magnetic mount antenna that you can quickly place on the roof.
Panic alarm	Use a personal panic alarm or whistle to call for help.
Tires	Make sure your spare is not flat and you have all the tire change equipment including a manual tire pump. Also include a quick tire repair kit (you spray into the valve and it seals from the inside)
Windup flashlight	Handy if out at night or in a tunnel.

Emergency kit	Make sure you have a complete emergency kit with blanket and disposable rain coat.
Jerry can	Keep a spare in the trunk. Empty and refill every month to keep fuel fresh.
Toolbox	For essential emergency repairs.
Swiss Army knife	Keep a couple of larger ones in your glove compartment and take with you when you leave your vehicle.

Preparation Tips for the office

Know the layout and policies	Familiarize yourself with your company's guidelines and remember where the assembly areas are across the building so you can go to the nearest one.
Food and bottled water	Make available vending machine water and food. Management should ensure food is equally distributed.
Emergency kit	Even the smallest office needs a first aid kit. Assemble a dedicated kit using first aid and medical, two-way radios, dosimeters and a Geiger counter or similar.
Lined metal box and radios	Keep one by your desk or work area so you can throw in your phone to protect against EMP. Keep a crank radio inside and a CB radio.

Preparation Tips for Your School

Know the layout and policies	Familiarize yourself with your school's guidelines and remember where the assembly areas are across the campus so you can go to the nearest one.
Water supply and food	There should be a dedicated, on-campus water supply. Keep a spare bottle in your locker, plus snack bars.
Panic alarms	Keep a whistle or panic alarm in your locker or on your person.
Emergency kits and Geiger counters	Ensure your emergency kit has enough two-way radios to pass around teachers and students. Assign to this kit a Geiger counter from the Science lab or similar, along with a number of dosimeters for Staff who may need to go outside.
Iodine tablets*	Discuss with your doctor if you need to have these as there are risks involved.

* It is not recommended to take iodine tablets unless authorized by a professional.

Follow the CDC guideline here,
https://www.cdc.gov/nceh/radiation/emergencies/ki.htm

Fact sheet on usage:
https://www.health.ny.gov/environmental/radiological/potassium_iodide/fact_sheet.htm

Notes

Notes

Notes

Disclaimer

The information in this book covers general emergency procedures which may differ from state-to-state. This book is sold with the understanding that the author and publisher are not providing medical advice or condoning the use of firearms. No information in this book is intended to replace any medical advice you may have. Weapons are at your sole risk and you must be licensed to use firearms. Check with your State on laws pertaining to weapons.

We make no representations or warranty expressed or implied on the completeness or accuracy or reliability with respect to information in this book. We have endeavored to provide the best quick-actionable advice and information; however the reliance you place in this book is at your own risk.

This book may be updated from time to time as preparation advice may change. We encourage you to seek further information on nuclear attack preparedness as it is a lengthy but worthwhile subject to learn about, in the interests of your safety and that of your family.

www.ingramcontent.com/pod-product-compliance
Lightning Source LLC
Chambersburg PA
CBHW071124240526
45465CB00023B/805